GROWING SUNFLOWER

The Ultimate Guide
on Sunflower
Cultivation and Care

Alex J. Morgan

Table of Contents

CHAPTER ONE

SUNFLOWERS

Composite flowers are essentially a collection of miniature flowers (florets) that when combined, give the impression of being one large bloom. Thus, the flower head we refer to as a "sunflower" is actually composed of countless smaller blooms. Numerous disk florets make up the flower's typically dark center. They'll eventually be pollinated, resulting in the production of sunflower seeds. The petals that make up the outer ring of the flower head are really ray flowers, which are sterile blooms meant to draw pollinating insects.

The sunflower, as their name suggests, is also a "sun tracker." Heliotropism is the botanical term for a plant's tendency to follow the sun. The plant can now get up to 10% more sunlight as a result! More sunlight is being reflected, causes increased plant growth, which is probably why these plants grow so quickly.

In addition to being stunning, sunflowers have a long history and a variety of purposes.

History of sunflower

One of the few domesticated natural plants in North America is the sunflower. The domestication of the sunflower is particularly unusual because most crop species were domesticated in places like the fertile crescent (ancient Mesopotamia), Asia, South America, and Central America.

Native Americans used prehistoric sunflowers, which were much smaller than their counterparts today, for food, fiber, and fuel. The stalks were used for construction, and the seeds were harvested for consumption.

The petals are used as pigments for dyes and other parts of the plant are used for medicinal purposes. The spread of sunflowers across North America may have been accidental, with the seeds accidentally landing on many Native American campsites. At that time, this plant was widely cultivated. The history of sunflower cultivation in eastern North America is also consistent with widespread sunflower cultivation in the southwestern United States and Mexico. Human mutation and selection over time has changed sunflowers. These plants have been bred to produce larger buds and most importantly larger seeds! Seeds from domesticated sunflowers are four times larger than the wild ancestral varieties of sunflowers. Sunflower breeding in the 20th century also resulted in earlier ripening, higher yielding varieties and semi-dwarf varieties. All these improved traits will lead to worldwide sunflower production. There are few flowers that have as big of an impact as the bright, cheerful and spectacular sunflower.

As a florist farmer. There are more than 150 species of sunflowers, but the common sunflower you see growing in most gardens is the giant sunflower Helianthus annuus. These beautiful flowers, along with marigolds, dandelions, lettuce and chrysanthemums, are part of the Asteraceae family. This family was formerly known as Compositae due to their characteristic of being composite flowers. Culture U.S. sunflower production accelerated in the Great Plains states following the reintroduction of new varieties. Sunflowers are grown for their oil seeds, bird seeds, and as a snack for humans. Sunflower acreage in the United States peaked in the 1970s, then declined due to crop prices and pest problems. Production has begun to recover gradually since then. Today, sunflowers are grown for the production of vegetable oils, snacks, bird seeds, and fodder. Feed is derived from sunflower seed parts left over from the oil extraction process, such as sunflower meal or sunflower pods. Sunflower meal contains 28-32% protein and is therefore a complete source of nutrients. If the seed has no shell or partial shell, the percentage of total protein will decrease. Sunflower shells are sometimes used as forage for livestock, but due to transportation costs, sunflower shells are often burned at oil extraction facilities. silage food Sunflowers can also be cut for silage, with the crude protein level of sunflower silage higher than corn silage but lower than alfalfa. The silage is created when the crop is cut at the "green" stage and subjected to anaerobic (no oxygen) conditions. Today, this fermentation is usually done through the use of silos or plastic bags. Sunflowers should be cut for silage when the first half of the flower contains immature seeds. Plants must be allowed to wilt and dry before silage. Silage is used as animal feed, especially cattle. Sunflower oil can also be used for industrial purposes, but since it is also considered a cooking oil, its price is much higher than that of petroleum raw materials or petroleum-based products. Sunflower husks have also been marketed as poultry litter, firewood, and other high-fiber specialty products. Sunflowers are also grown as a special cut flower for flower arrangements and weddings.

CHAPTER TWO

PROPAGATION

First of all, if you're looking for a beautiful and easy-to-grow flower in your garden, it's sunflowers! Seeds are usually fairly inexpensive and can be easily found at your local grocery or hardware store. And speaking of sunflower seeds, seeds are the best way to propagate sunflowers! If you've ever bought a bag of sunflower seeds, you're probably familiar with the black and white or gray and white outer shells of sunflower seeds. However, sunflower seeds can also be black! Some are small and some are larger, it just depends on the breed. start inside Sunflowers can be started indoors or planted directly in the garden. If you're starting to grow indoors, sow the seeds about 4 to 6 weeks before your estimated final frost. Make sure your seeds are sown 1 inch deep. The general rule of seeding depth is to bury the seed at least 1-2 times the seed size. Keep your pot or seed tray moist while the seeds begin to germinate. Once the seedlings have sprouted, be sure to water them when the seeding mix begins to dry out. Sow directly in the garden, to sow seeds in a garden or landscape, use a triangular hoe or hand shovel to create a trench. Place the seeds about 6 to 12 inches apart. Some professional cut flower growers prefer to plant their sunflowers 1 to 2 inches apart, as the sunflower heads will be smaller and better suited for bouquets.

If you want the top of your sunflower to be taller, make sure to provide plenty of room for the plant, especially branched sunflowers. Cover the trench with at least 1 to 2 inches of soil. Some people are very lucky to scatter sunflower seeds on the surface of their soil. Be aware that seeding often results in a lower germination rate, but at least you'll have a few plants growing with your efforts anyway! In fact, in some areas, sunflowers can be considered a weed due to their ability to self-seed. Remember that seeds from a hybrid sunflower will not look like the parent sunflower. They will likely serve their purpose in your garden no matter what they look like.

When to plant sunflower

Sunflowers should be planted or transplanted after last spring's frosts according to your estimates. Occasionally, some varieties will survive a light frost, but it's best to avoid planting too early if possible. Mature sunflowers can withstand temperatures of about 25 degrees F. Seeds can germinate when soil temperatures are as low as 46 degrees F. Sunflowers can be planted consecutively until midsummer for a fresh source of flowers. If you want to know when your last sunflower planting was, find the estimated first frost date in fall and the average number of days it takes for your sunflower variety to mature. Then, using your average days to maturity, count down from your first frost and you'll have some time to plant your last batch of sunflowers!

CHAPTER THREE

HOW TO GROW
SUNFLOWER

As the name suggests, sunflowers should be planted in full sun (more than 6-8 hours of direct sun per day). Sunflowers prefer soils rich in moisture and nitrogen, but can be grown in almost any soil from high sand content to high clay content. Floor They prefer soil with a pH between 6.0 and 6.8. They are inefficient and patriotic water users, but can handle some drought. For these reasons, sunflowers can handle small amounts of stress, but the most important times to keep your sunflowers stress-free are during the 20 days before and 20 days after flowering. Amending the soil with compost will help increase your organic matter content, and thus increase the water holding capacity and nutrient holding capacity of the soil. plant the location of the sunflowers in the garden is important as most varieties will be several feet tall. Their height can be detrimental to nearby plants that may need full sun to grow. Plant your sunflowers at the bottom of a garden or a few feet away from shorter plants that need full sun. You can use the height of the sunflower to your advantage for semi-sun-loving plants. watering needs Water your sunflowers well as they begin to grow to promote a healthy root system. Sunflowers may not need staking if they have established a strong root system. However, in windy areas, they can benefit from staking. Before fertilizing, it is best to conduct a soil test for the fertility of a particular soil type before amending it. Sunflowers can benefit from balanced fertilization (that is, a fertilizer that contains equal parts nitrogen, phosphorus, and potassium). Another alternative that is considered organic is to use a mixture of fish emulsion and liquid seaweed. This can be applied every few weeks through foliar application or applied by watering can to the soil. Always follow the manufacturer's instructions on mixing ratios and fertilizer use. After your sunflowers take off, they probably won't need to fertilize. group Depending on the variety of sunflower you are growing; you may want to prune your sunflower when it reaches a height of at least 12 inches. Sunflowers that are considered branched will produce buds in the leaf axils that will develop into flowers, resulting in a plant with many blooms. branch out When young, you can pluck the branching sunflowers just above the leaves, this will encourage the plant to produce more flowers on long, straight stems suitable for cutting. If left whole, branching sunflowers will also produce many branches, but they are usually shorter and unsuitable for cutting due to their short stem length. It really depends on the home gardener and their goals for their sunflower plants! The unbranched sunflower has been specially bred to produce a large, uniform flower.

When and how to harvest

You will harvest your sunflowers at different stages depending on your growing goals! If you want to use them as cut flowers, you can cut them any time after the petals begin to lift off the center plate. Sunflowers will continue to open in the vase after being cut. Cut the base of the flower stalk for a branched sunflower, or to whatever length you want if it's an unbranched sunflower. I cut the stems as long as possible so I can cut them to the right length for my vase when I put them in the vase. If you want to harvest sunflowers for seeds, wait until the flowers fade. Usually, the back of the flower head turns yellow or brown when the seeds are fully ripe and the flower heads dry. You will notice that the top begins to drop slightly and the center disc becomes rounder and fuller as the seeds have grown. Remember to collect the seeds before the birds get there if you want to save them! The easiest way I've found to collect seeds is to cut off the tip and stir the seeds with your fingers in a bowl or on a cloth. If you want to feed your birds with sunflower seeds, you should cut off two or three sunflower heads after the seeds have grown and bundle them together. Hang this set upside down on a porch pole or anywhere you like for bird watching. Sunflower heads become dry decorations and all-in-one bird feeders! 45 species of birds are known to eat sunflower seeds.

Sunflower varieties

Types of sunflowers Varieties are grouped into several categories, including giant, dwarf, non-pollen, branched, and unbranched. Branching sunflowers will produce multiple sunflowers per plant, which is why they are garden favorites! Unbranched sunflowers are sunflowers (usually hybrids) that have been specially bred to have a large, uniform head and early ripening date. They are a favorite of cut flower farms and oilseed growers. Some varieties of sunflowers don't even have pollen, so if you cut those sunflowers to make a bouquet, you won't have all the yellow pollen falling on your white tablecloth. Your seed catalog or seed company will usually tell you whether the variety is branched or unbranched. branch-like; Sonja – Dark disc-shaped citrus flowers, ideal for cutting, about 42 inches tall, Shock o` Lat - F1 hybrid sunflowers, dark chocolate winged pollenless flowers with bright yellow tips, great for cut flowers, Joker - F1 hybrid sunflower grows to 6 feet tall with delightful semi-double crest flowers 4 inches in diameter, flowers are bicolor mahogany with yellow tips and are pollen-free. This plant has a short ripening period of 55 days, Teddy Bear - Dwarf cotton 16 inches tall, double flowers yellow, Moulin Rouge or Rouge Royale - Chalk-free sunflower with delicate dark red flowers. unbranched variety *Procut Line - This strain is pollen free, day free, early ripening (55 days) and is excellent for producing cut flowers with uniform 3-4 inch blooms. Colors include orange, lemon, yellow, two-tone, plum, red, white, and more.

Some varieties have clear discs rather than brown ones. *Sunrich Series - This strain is about 3 to 4 feet tall and pollen free with flowers that are 5 to 6 inches in size on average. The flowers ripen within 55-70 days from sowing. The colors orange, lemon yellow, lemon and yellow (with green center). *Double Quick - 5 feet tall within 5-inch long sunflower clusters, distinctly concentrated on the grass. Pollen-free varieties mature about 65 days after sowing. *Vincent`s Choice - 5-6 foot tall variety with overlapping orange petals making it look like a semi-double flower. Great cut flowers to transport and arrange. giant breed The flowers are large, dark yellow, and up to 4 to 6 inches in diameter.

CHAPTER FOUR

PEST AND DISEASE CONTROL

Sunflowers can be severely affected by disease and insect pests. There are several farming methods that will help reduce the incidence of diseases and pests in your garden, which will be discussed below. Common sunflower disease Most diseases that affect sunflowers are fungal and affect the lower leaves of the plant first, so if the plant is mature when infected, the disease may not be a problem.

Sclerotinia stem rot (Verticillum dahliae) causes wilting soon after flowering and a light brown stripe at the base of the stem. Seeds and seed flesh will be discolored. Rust (Pucchini helianthi) causes rust-colored pustules on leaves and black spots on stems. Leaf spot disease will cause dead spots on the leaves. Late blight (Plasmopara hastedi) causes cottony fungal growth on the undersides of leaves and stunting and/or discoloration, while powdery mildew (Erysiphe cichoracearum) causes cottony growth on common leaves. at the end of summer. Crop rotation in your garden will help limit disease in your sunflowers. Crop rotation is the practice of rotating plants or plant families in your garden to help prevent the accumulation of disease or pest pressure. Extending crop spacing will also help control diseases such as powdery mildew who like wet conditions. Leaving more space between plants will increase air flow between plants to ensure rain or dew evaporates quickly from the leaves. Watering the plant at the base rather than at the top will also help prevent disease incidence and spread, as will planting in a well-drained location, as most diseases are spread by interacting with water. Growing disease-resistant varieties is also a great cultural practice. The seed directory or seed company will often have information on the diseases to which each variety may be resistant. Over time, plant breeders have selected plant varieties that are naturally resistant to certain diseases, and they use these plants when crossbreeding to create new varieties, creating new varieties. Healthy, uniform and disease-resistant progeny can be grown as commercial crops. Farmer and hobby gardener. Common Sunflower Pests The larvae of sunflower moths, sunflower moths and sunflower moths feed on the flower heads, stems, and seeds of sunflowers. Usually, symptoms of larval feeding will be digging holes in the stem or head of the sunflower. This can lead to flower head deformation or seed drop. Sunflower pest cause flower rays (petals) to be lost or flower heads cut off. Sunflower maggots burrow into tree trunks.

Beetle larvae are yellow, rough larvae that cause severe defoliation. Stem borers and cup weevils can cut off flower heads or severely damage leaves. Because an economic threshold for pesticide use on sunflowers has not been established and because of its ability to kill beneficial pollinators, it is not advisable to spray pesticides on sunflowers. If insect larvae or eggs are present on sunflower stems and leaves, one option is to remove them from the plant. Releasing beneficial insects into your garden, which feed on caterpillars or eggs, and crop rotation are two other ways to reduce insect pressure. Examples of beneficial insects are ladybugs, mantises, and beetles. Birds and deer are two non-insect pests of sunflowers. Birds like to eat sunflower seeds, while deer prefer to eat immature sunflower plants. Use fences or repellents to control deer feeding. Artificial owl lures, gyros and scarecrows will help deter birds from eating seeds if your goal is to harvest sunflower seeds for use.

Preservation

To save sunflower seeds for bird feeding or snacking, cut off the flower heads at the right time of harvest and let them dry. Seeds should not be moist during storage. They should also be stored in an airtight container or paper bag in a cool, dry place. If you keep the flower heads for bird feeding in the winter, after drying the heads in a warm and ventilated place, simply place the flower heads in a paper bag. Store them in a dry place until you are ready to take them out to feed the birds. Sunflowers can be cut for fresh or dried flower arrangements. Fresh sunflowers can be cut as soon as the petals begin to lift and open and stored in the refrigerator for long term storage. Sunflower heads can also be dried by hanging them upside down in a warm, dry, and well-ventilated space. However, the petals will shrivel and deform. Another option is to bury the flower head in silica sand for several weeks in an airtight container. The silica sand will absorb moisture from the flower head. To harvest sunflowers for drying, make sure the flowers are fully open. Uses of the tree Commercially, sunflower seeds are used to make oil, animal feed and human snacks, as well as many other special uses. Historically, sunflowers were used by Native Americans for food, fuel, and construction purposes. These plants are also celebrated for their large, beautiful flowers in gardens and fresh flower arrangements. Sunflower fields are a popular photo opportunity for photographers and agritourism farms. They attract pollinating birds and insects, so they are a great addition to a garden or array of pollinators-friendly vegetables.